KRISHNA MOHAN
AVANCHA

Rising From Ashes!!!

Contents

1

Chapter 1: Introduction

In the world of business, brands are constantly evolving and adapting to the changing needs of their customers. Some brands thrive, while others struggle to keep up with the competition. But what happens when a brand faces criticism and legal complaints from its customers? Can it rise from the ashes and regain the trust of its audience?

In this book, we will explore the concept of the phoenix brand – a brand that rises from the ashes of negative publicity and emerges stronger than before. We will examine the strategies and tactics used by these brands to win back their customers and rebuild their reputation.

Digital marketing has played a crucial role in the success of phoenix brands. Through social media, email marketing, search engine optimization, and other digital channels, these brands have been able to reach out to their customers and rebuild trust.

One of the most famous examples of a phoenix brand is Tylenol. In 1982, Tylenol faced a major crisis when seven people died after taking cyanide-laced capsules of the pain reliever. Tylenol quickly took action, recalling all products from store shelves and implementing tamper-proof packaging. The company also launched a major public relations campaign to rebuild its reputation. Today, Tylenol is one of the most trusted brands in the pain relief category.

Another example is Johnson & Johnson, the parent company of Tylenol. In 2010, the company faced a crisis when it was discovered that some of its products, including Tylenol, were contaminated with bacteria. The company quickly launched a recall and implemented new safety procedures. Johnson & Johnson also used social media and other digital channels to communicate with its customers and rebuild trust. Today, the company is still one of the most respected and trusted brands in the healthcare industry.

Other phoenix brands include Starbucks, which faced criticism over the closure of its stores for racial bias training in 2018, and Nike, which faced backlash over its Colin Kaepernick ad campaign in 2018. Both companies were able to use digital marketing to reach out to their customers and rebuild their reputations.

In this book, we will examine these and other examples of phoenix brands, exploring the strategies and tactics they used to overcome negative publicity and emerge stronger than before. We will also look at the role of digital marketing in the success of these brands, and how businesses can use digital channels to build and maintain trust with their customers.

The journey of a phoenix brand is not an easy one. It requires courage, transparency, and a commitment to change. But for those brands that are able to rise from the ashes of negative publicity, the rewards can be great. They can earn the trust and loyalty of their customers, and emerge as stronger, more resilient brands.

Some examples of real estate companies which rebuilt their brand:

- **Zillow**

Zillow is a popular online real estate marketplace that allows users to search for properties and connect with real estate agents. In 2017, the company faced criticism when it was discovered that its "Zestimate" feature, which estimates the value of homes, was often inaccurate. Zillow used digital marketing to address the issue, launching a campaign that focused on transparency and improving the accuracy of its estimates. The company also used social media to communicate with customers and respond to their concerns. Today, Zillow continues to be a popular platform for real estate search and has improved the accuracy of its estimates.

- **Redfin**

Redfin is another online real estate marketplace that connects buyers and sellers with agents. In 2014, the company faced a class-action lawsuit from a group of sellers who claimed that the company's commission structure was anticompetitive. Redfin used digital marketing to respond to the lawsuit, launching a social media campaign that highlighted its commitment to transparency and customer satisfaction. The company also used email marketing to communicate with customers and keep them updated on the lawsuit. The lawsuit was eventually dismissed, and Redfin continues to be a popular platform for real estate search.

- **Trulia**

Trulia is a real estate platform that allows users to search for homes, apartments, and other properties. In 2016, the company faced criticism when it was discovered that some of the listings on its platform were fraudulent. Trulia used digital marketing to address the issue, launching a campaign that focused on improving the accuracy of its listings and increasing transparency for users. The company also used social media to communicate with customers and respond to their concerns. Today, Trulia continues to be a popular platform for real estate search and has implemented new measures to prevent fraudulent listings.

- **Compass**

Compass is a real estate brokerage firm that uses technology to streamline the buying and selling process. In 2019, the company faced criticism from some agents who claimed that its technology-focused approach was putting traditional agents out of business. Compass used digital marketing to respond to the criticism, launching a campaign that highlighted its commitment to supporting agents and providing them with the tools they need to succeed. The company also used social media to communicate with agents and respond to their concerns. Today, Compass continues to be a popular platform for real estate search and has expanded its offerings to include more traditional services.

In each of these examples, digital marketing played a key role in helping real estate companies overcome challenges and improve their offerings. By using social media, email marketing, and other digital channels to communicate with customers and address concerns, these companies were able to rebuild trust and maintain their position as leaders in the real estate industry.

Chapter 2: Understanding the Problem

Before we can begin to explore the strategies and tactics used by phoenix brands to overcome negative publicity and legal complaints, we must first understand the problem that these brands are facing. In this chapter, we will examine the common causes of negative publicity and legal complaints, and the impact that they can have on a brand's reputation.

Causes of Negative Publicity and Legal Complaints

Negative publicity can stem from a variety of sources, including product defects, poor customer service, unethical business practices, and employee misconduct. When customers feel that they have been wronged or deceived, they may take to social media and other online channels to voice their complaints, which can quickly escalate and cause significant damage to a brand's reputation.

Legal complaints can also arise from a variety of issues, including false advertising, breach of contract, and product liability. When a brand is faced with a legal complaint, it can be a costly and time-consuming process that can further damage the brand's reputation and erode customer trust.

Impact on a Brand's Reputation

Negative publicity and legal complaints can have a significant impact on a brand's reputation. Customers who have had negative experiences with a brand are likely to share their experiences with others, potentially leading to a widespread loss of trust in the brand. Negative publicity and legal complaints can also damage a brand's credibility and make it difficult for the brand to attract new customers.

In addition to the direct impact on a brand's reputation, negative publicity and legal complaints can also have financial consequences. Brands may be required to pay settlements, fines, and legal fees, which can be a significant expense. They may also experience a decline in sales and revenue as a result of the negative publicity.

The Role of Digital Marketing in Addressing Negative Publicity and Legal Complaints

Digital marketing can play a crucial role in addressing negative publicity and legal complaints. Through social media, email marketing, and other digital channels, brands can communicate with customers and address their concerns in a timely and transparent manner. By being proactive in addressing negative publicity and legal complaints, brands can demonstrate their commitment to their customers and work to rebuild trust.

Digital marketing can also be used to highlight the positive aspects of a brand and its products or services. By creating engaging content that showcases the brand's strengths, brands can build a positive reputation and strengthen customer loyalty.

Conclusion

Negative publicity and legal complaints can have a significant impact on a brand's reputation and financial performance. In order to address these issues, brands must be proactive in communicating with customers and addressing their concerns. Digital marketing can be a powerful tool in this process, allowing brands to engage with customers in a transparent and timely manner. In the next chapter, we will examine the strategies and tactics used by phoenix brands to overcome negative publicity and legal complaints and emerge stronger than before.

Chapter 3: Assessment and Analysis

Assessment and analysis are critical components of any successful project or program. They allow organizations to understand their strengths and weaknesses, identify areas for improvement, and develop strategies for achieving their goals.

Assessment refers to the process of gathering information about a project, program, or organization. It can involve collecting data through surveys, interviews, focus groups, or other methods. The goal of assessment is to gain a clear understanding of the current situation and identify any gaps or areas for improvement.

Analysis, on the other hand, involves examining the data collected during the assessment process to draw conclusions and make recommendations. This can involve statistical analysis, qualitative analysis, or a combination of both.

Assessment and analysis are important for several reasons. First, they help organizations to identify their strengths and weaknesses. By understanding what they do well and where they need to improve, organizations can develop strategies to address their weaknesses and build on their strengths.

Second, assessment and analysis help organizations to set goals and develop strategies for achieving those goals. By understanding the current situation and identifying areas for improvement, organizations can develop specific, measurable, achievable, relevant, and time-bound (SMART) goals that will guide their efforts.

Third, assessment and analysis help organizations to track their progress and evaluate their success. By monitoring their progress over time, organizations can identify what is working well and what needs to be improved.

To conduct an effective assessment and analysis, organizations should follow a structured process. This process typically involves several

steps, including:

Defining the purpose and scope of the assessment. This involves identifying the specific questions that the assessment will seek to answer and the population that will be surveyed.

Selecting assessment methods. This involves selecting the appropriate data collection methods, such as surveys, interviews, or focus groups.

Designing assessment instruments. This involves developing the actual survey or interview questions that will be used to collect data.

Collecting data. This involves administering the assessment instruments to the target population and collecting the data.

Analyzing the data. This involves reviewing the data collected and identifying key trends and patterns.

Reporting the findings. This involves communicating the results of the assessment and analysis to key stakeholders and using those results to inform decision-making.

Assessment and analysis are critical components of any successful project or program. By gathering data, identifying strengths and weaknesses, and developing strategies for improvement, organizations can achieve their goals and drive meaningful change.

4

Chapter 4: Communicate and Apologize

Effective communication is crucial in resolving conflicts and repairing relationships. When we communicate in a clear, respectful, and empathetic manner, we increase the chances of finding a mutually satisfactory solution. Apologizing, when necessary, can also play a crucial role in the process of conflict resolution and relationship repair. In this chapter, we will explore the importance of effective communication and apology in the context of conflict resolution.

Effective Communication
Effective communication involves the exchange of information, thoughts, and feelings in a way that is clear, respectful, and empathetic. When we communicate effectively, we are able to express our

own needs and concerns while also listening to and understanding the needs and concerns of others.

Here are some key elements of effective communication:

Active Listening: Active listening involves fully concentrating on what the other person is saying, without interrupting or judging. It is important to ask clarifying questions and restate what the other person has said to ensure that you have understood them correctly.

Use "I" Statements: Using "I" statements can help to avoid blame and defensiveness. For example, saying "I feel hurt when you interrupt me" is more effective than saying "You always interrupt me and it's so annoying."

Focus on the Present: When discussing a conflict or disagreement, it's important to focus on the present situation rather than bringing up past grievances or issues. This can help to avoid escalating the conflict.

Stay Calm: It's important to remain calm and composed during communication, even if the other person becomes angry or defensive. Taking deep breaths, using relaxation techniques, or taking a break if necessary can help to maintain a calm and respectful tone.

Empathy: Showing empathy involves understanding and acknowledging the other person's feelings and perspective. This can help to build rapport and create a more positive atmosphere for communication.

Apologizing

Apologizing can be a difficult but important step in the process of conflict resolution and relationship repair. A sincere apology can help to acknowledge wrongdoing, take responsibility, and express regret for any harm caused. Here are some key elements of a sincere apology:

Acknowledge the Hurt: Begin by acknowledging the harm that was caused and the impact it had on the other person. For example, "I understand that my actions hurt you deeply."

Take Responsibility: Accept responsibility for your actions and any consequences that resulted. Avoid making excuses or blaming others.

Express Regret: Express genuine remorse for the harm caused, and acknowledge the impact it had on the other person. For example, "I am truly sorry for the pain I caused you."

Offer Amends: Offer to make things right, if possible. This may involve making restitution or taking steps to prevent the harm from happening again.

Commit to Change: If the harm was caused by a behavior or pattern of behavior, commit to changing that behavior and taking steps to prevent similar situations in the future.

Conclusion

Effective communication and apology are essential components of conflict resolution and relationship repair. When we communicate in a clear, respectful, and empathetic manner, we increase the chances of finding a mutually satisfactory solution. Apologizing can help to acknowledge wrongdoing, take responsibility, and express regret for any harm caused. By utilizing these skills, we can build stronger, more positive relationships and resolve conflicts in a constructive manner.

5

Chapter 5: Addressing the Issues

A phoenix brand is a brand that has experienced a crisis or downturn, often resulting in significant negative publicity, lost market trust, and reduced consumer confidence. This chapter will examine the challenges faced by phoenix brands and discuss strategies for addressing these issues in order to successfully rebuild and regain market trust.

Identifying the Issues:

The first step in addressing the problems faced by a phoenix brand is to identify the root causes of the crisis. This may include:

Product or service failures

Management issues
Unethical business practices
Negative publicity

Regulatory or legal problems

By pinpointing the specific issues at hand, the brand can then develop targeted strategies to address and overcome them.

Apologizing and Taking Responsibility:
An essential step in rebuilding trust is to issue a public apology, admitting fault and taking responsibility for the mistakes made. This demonstrates transparency, accountability, and a commitment to learning from past errors.

Implementing Changes:
To regain market trust, the phoenix brand must make substantive changes in their operations, products, or services. This may involve:

Revising business practices to align with ethical standards
Overhauling management structures
Improving product quality and safety
Enhancing customer service
Updating branding and messaging

Communicating with Stakeholders:
Transparent and open communication is crucial in rebuilding trust. A phoenix brand should con-

sistently update stakeholders, including customers, employees, and investors, on the progress being made in addressing the issues. This can be achieved through press releases, social media updates, and direct communication channels.

Engaging with the Community:

Rebuilding a positive brand image involves engaging with the community through charitable initiatives, sponsorships, or other forms of corporate social responsibility (CSR). This helps the brand demonstrate its commitment to bettering itself and giving back to society.

Monitoring and Evaluation:

To ensure that changes are effective and sustainable, phoenix brands should establish systems for monitoring and evaluating their progress. This may involve conducting regular audits, soliciting customer feedback, or utilizing third-party evaluators. This data should be used to make ongoing improvements and maintain transparency with stakeholders.

Celebrating Successes:

As the phoenix brand makes progress in addressing its issues, it is important to celebrate milestones and communicate these achievements to stakeholders. This reinforces the message that the brand is actively working to overcome its challenges and helps rebuild trust.

Story Example 1:
Once upon a time, in a small town, there was a popular bakery called "Delightful Bites." The bakery was famous for its delicious pastries and cakes, and it had gained the trust and loyalty of the local community. However, a sudden turn of events changed everything.

Identifying the Issues:
One day, news broke out that Delightful Bites had been using poor-quality, expired ingredients in their products. This led to several customers falling ill, and the bakery's reputation was tarnished overnight. The bakery lost a significant portion of its customer base and faced a crisis. The owner, Mrs. Brown, knew she had to identify the root cause of the problem and take action.

Apologizing and Taking Responsibility:
Mrs. Brown quickly issued a heartfelt public apology, taking full responsibility for the use of expired ingredients. She promised her customers that she would rectify the situation and regain their trust.

Implementing Changes:

To make amends and rebuild her bakery's reputation, Mrs. Brown immediately implemented several changes. She hired a quality control expert, invested in better storage facilities, and ensured that all ingredients were sourced from reputable suppliers. She also revamped the bakery's menu, focusing on healthier and higher-quality products.

Communicating with Stakeholders:

Throughout the process, Mrs. Brown remained transparent about the steps being taken to improve the bakery. She sent regular newsletters to customers, shared updates on social media, and held community meetings to address concerns and answer questions.

Engaging with the Community:

To demonstrate her commitment to the community, Mrs. Brown initiated a "Baking for a Cause" program. Delightful Bites partnered with a local charity, donating a portion of their proceeds and organizing free baking workshops for underprivileged children. This initiative helped to rebuild the bakery's positive image.

Monitoring and Evaluation:

Mrs. Brown established a system to monitor and

evaluate the bakery's progress. She collected customer feedback through surveys, regularly reviewed quality control reports, and hired an external auditor to ensure compliance with industry standards. This allowed her to make continuous improvements and maintain transparency with her customers.

Celebrating Successes:

As Delightful Bites began to regain its customers and rebuild its reputation, Mrs. Brown celebrated the milestones along the way. She hosted a "Customer Appreciation Day," offering free samples of their new and improved pastries, and shared positive customer testimonials on social media.

Story Conclusion:

Through hard work, transparency, and a commitment to change, Mrs. Brown and Delightful Bites were able to rise from the ashes, overcoming their crisis and regaining the trust of the community. The bakery's journey serves as an example of how a phoenix brand can successfully address its issues and come back stronger than before.

Story Example 2:

Once upon a time, in a bustling city, there was a renowned real estate company called "Dream Homes Inc." The company was known for its luxurious and high-quality residential projects, and it had earned the trust and loyalty of its clients. However, a series of unfortunate events led to a severe blow to the company's reputation.

Identifying the Issues:

News broke that Dream Homes Inc. had been cutting corners in construction, leading to safety hazards and code violations in their properties. As a result, residents faced numerous issues, such as water leaks, faulty wiring, and structural weaknesses. The company's reputation plummeted, and many clients chose to terminate their contracts. The CEO, Mr. Smith, knew he had to identify the root cause of the problem and take action.

Apologizing and Taking Responsibility:

Mr. Smith promptly issued a sincere public apology, taking full responsibility for the compromised construction quality. He assured clients that the company would rectify the situation and regain their trust.

Implementing Changes:

To make amends and rebuild the company's reputation, Mr. Smith implemented several changes. He hired a team of independent inspectors to assess all ongoing projects, and he brought in a new construction manager to ensure strict adherence to safety standards and building codes. He also committed to using only high-quality materials and employing skilled labor for all future projects.

Communicating with Stakeholders:

Throughout the process, Mr. Smith remained transparent about the steps being taken to improve Dream Homes Inc.'s operations. He sent regular updates to clients, shared progress reports on the company's website, and held open forums to address concerns and answer questions.

Engaging with the Community:

To demonstrate the company's commitment to the community, Dream Homes Inc. initiated a "Homes for All" program. The company partnered with a local nonprofit organization to build affordable housing units for low-income families, using the same quality materials and construction standards as their luxury projects. This initiative helped to rebuild the company's positive image.

Monitoring and Evaluation:

Mr. Smith established a system to monitor and evaluate the company's progress. He collected feedback from clients and residents, reviewed inspection reports, and commissioned third-party audits to ensure compliance with industry standards. This allowed him to make continuous improvements and maintain transparency with his clients.

Celebrating Successes:

As Dream Homes Inc. began to regain its clients and rebuild its reputation, Mr. Smith celebrated the milestones along the way. He hosted a grand opening for the company's latest luxury development, showcasing the improved construction quality, and shared positive client testimonials in their marketing materials.

Story Conclusion:

Through hard work, transparency, and a commitment to change, Mr. Smith and Dream Homes Inc. were able to rise from the ashes, overcoming their crisis and regaining the trust of their clients. The company's journey serves as an example of how a phoenix brand can successfully address its issues and come back stronger than before in the competitive real estate market.

Conclusion:

Addressing the issues faced by a phoenix brand is a challenging but necessary process. By identifying the root causes of the crisis, taking responsibility, and implementing targeted changes, a phoenix brand can successfully rebuild and regain market trust. Transparent communication, community engagement, and ongoing monitoring and evaluation are key components of this journey, ultimately leading to a stronger and more resilient brand.

Chapter 6: Encouraging Positive Reviews

When a brand has been criticized and lost market trust, it can be challenging to encourage positive reviews. However, it is not impossible, and there are several steps that a phoenix brand can take to regain its reputation and encourage positive reviews.

Address the Issues: The first step in encouraging positive reviews is to address the issues that led to the criticism in the first place. The brand needs to identify the root cause of the problem and take corrective action to address it. This could involve changing the product or service, improving customer service, or changing the company culture.

Engage with Customers: It is essential to engage with customers and listen to their feedback. The brand should be responsive to customer complaints and take the necessary steps to address them. By engaging with customers and addressing their concerns, the brand can show that it is committed to improving its products and services.

Offer Incentives: Offering incentives such as discounts or free products can encourage customers to leave positive reviews. The brand could offer incentives for customers who leave a review on a particular platform or who share their experience on social media. By offering incentives, the brand can encourage customers to share their positive experiences with others.

Leverage Influencers: Influencers can be a powerful tool for promoting a brand and encouraging positive reviews. The brand could partner with influencers who have a strong following on social media and who are willing to promote the brand to their followers. By leveraging influencers, the brand can reach a wider audience and increase its visibility.

Focus on the Positive: Finally, it is essential to focus on the positive aspects of the brand. The brand should highlight its strengths and showcase its successes. By focusing on the positive, the brand can shift the conversation away from the negative and encourage customers to leave positive reviews.

Story Example:
Once a popular real estate brand, Atlas Real Estate had been struggling to maintain its reputation after a recent controversy. Some of their clients had complained about their lack of transparency and poor communication. As a result, the company had lost several clients, and their online reviews had suffered. Atlas Real Estate realized they needed to take action to restore their reputation.

Address the Issues: The first step Atlas Real Estate took was to address the issues that led to the criticism. They implemented new policies and procedures to increase transparency and improve communication with clients. They also ensured that their agents were trained to handle client concerns promptly.

Engage with Customers: Atlas Real Estate reached out to their past clients and asked for feedback on their experience. They listened carefully to the concerns and complaints of their clients and made sure to respond to each one personally. This demonstrated that the company was committed to improving and that they valued their clients' feedback.

Offer Incentives: To encourage positive reviews, Atlas Real Estate offered a discount on their services to clients who left a review on Google, Yelp, or their social media platforms. This incentivized clients to leave positive feedback and helped to improve the company's online reputation.

Leverage Influencers: Atlas Real Estate partnered with several influencers in the real estate industry who had a strong following on social media. These influencers shared positive stories about their experience with Atlas Real Estate, which helped to build trust and credibility with their followers.

Focus on the Positive: Finally, Atlas Real Estate focused on the positive aspects of their business, such as their high-quality properties, experienced agents, and exceptional customer service. They shared success stories and positive reviews on their website and social media platforms to showcase their strengths and achievements.

Story conclusion, by addressing the issues, engaging with customers, offering incentives, leveraging influencers, and focusing on the positive, Atlas Real Estate was able to regain its reputation and encourage positive reviews. The company's efforts paid off, and they were able to attract new clients and build a stronger, more loyal customer base.

In conclusion, encouraging positive reviews for a phoenix brand that has been criticized and lost market trust can be challenging, but it is not impossible. By addressing the issues, engaging with customers, offering incentives, leveraging influencers, and focusing on the positive, the brand can regain its reputation and encourage positive reviews.

7

Chapter 7: Building a Strong Brand Reputation

In this scenario, we can use the metaphor of a phoenix rising from the ashes to describe the process of rebuilding a brand that has faced significant challenges.

The first step in rebuilding a brand is to acknowledge the problems that led to the loss of trust in the first place. This requires a deep understanding of the reasons why customers, stakeholders, or the media may have criticized the brand. Once these issues are identified, the brand can begin to address them through a combination of communication and action.

Communication is a critical part of rebuilding a brand reputation. The brand must be transparent about what went wrong and what steps it is taking to address the issues. This may involve issuing public apologies, communicating with customers and stakeholders through social media or other channels, and being responsive to feedback.

In addition to communication, the brand must also take concrete steps to address the issues that led to the loss of trust. This may involve improving product quality, changing business practices, or addressing any ethical concerns that were raised.

One effective strategy for rebuilding a brand is to focus on a core set of values that the brand can embody and communicate to customers. These values should be aligned with the needs and desires of the target audience and should be reflected in all aspects of the brand, from product design to marketing communications.

Another strategy is to create a sense of momentum around the brand by launching new products, services, or marketing campaigns that demonstrate the brand's commitment to its core values. This can help to re-engage customers and stakeholders and create positive momentum around the brand.

Ultimately, rebuilding a brand after it has faced criticism and lost market trust requires a combination of communication, action, and a clear focus on core values. By acknowledging the issues that led to the loss of trust and taking concrete steps to address them, a brand can begin to rebuild its reputation and regain the trust of its customers and stakeholders.

Story Example:

In the early 2000s, a real estate development company named "Greenfields" was known for building high-end, luxury properties in prime locations. However, in 2008, during the global financial crisis, Greenfields faced significant financial challenges and was forced to cut corners on some of its developments, resulting in poor quality construction and delays in project completion. As a result, Greenfields' reputation suffered, and the company lost the trust of its customers and stakeholders.

To address these challenges, Greenfields took a number of steps to rebuild its brand reputation. First, the company acknowledged the problems it faced and publicly apologized for any shortcomings in its developments. The company then launched a comprehensive program to improve the quality of its construction, including investing in better materials and hiring more experienced contractors. Greenfields also implemented a more rigorous project management process to ensure that projects were completed on time and to a high standard.

In addition to these operational changes, Greenfields also launched a new marketing campaign that focused on the company's core values of quality, reliability, and customer service. The campaign featured testimonials from satisfied customers, highlighting the company's commitment to delivering high-quality properties and outstanding customer service.

Over time, Greenfields was able to rebuild its brand reputation and regain the trust of its customers and stakeholders. The company continued to focus on its core values and invested in new technologies and sustainable development practices, which helped to differentiate it from competitors and establish it as a leader in the industry.

In conclusion, the example of Greenfields shows that it is possible for a real estate brand to rebuild its reputation after facing significant challenges. By acknowledging the problems it faced, taking concrete steps to address them, and communicating its core values and commitment to quality, Greenfields was able to regain the trust of its customers and stakeholders and establish itself as a leading real estate developer.

8

Chapter 8: Engaging with Customers

A phoenix brand is a term used to describe a brand that has successfully risen from the ashes after being criticized and losing market trust. The term is derived from the mythical bird, the phoenix, which is known for being reborn from its own ashes. Engaging with customers is a critical step in the process of rebuilding trust and re-establishing a brand's reputation.

Here are some steps that a company can take to engage with customers after being criticized and losing market trust:

Acknowledge the problem: The first step in engaging with customers is to acknowledge the problem. The company needs to admit that there was an issue and take responsibility for it. This can be done through a public apology or a statement on the company's website.

Listen to customers: Listening to customers is essential in rebuilding trust. The company needs to be open to feedback and suggestions from its customers. This can be done through surveys, social media listening, or customer support channels.

Take action: Once the company has acknowledged the problem and listened to its customers, it needs to take action to address the issue. This can involve changes to the company's processes, policies, or products. The company should communicate these changes to its customers and demonstrate that it is taking steps to prevent the problem from happening again.

Engage with customers: Engaging with customers is an ongoing process. The company should continue to listen to feedback and suggestions from its customers and make changes accordingly. The company can also engage with customers through social media, email newsletters, or other marketing channels.

Rebuild trust: Rebuilding trust takes time, but it is

possible. The company needs to be transparent and honest with its customers. It should communicate regularly with its customers and demonstrate that it is committed to making things right. The company can also offer incentives or promotions to encourage customers to give it another chance.

In summary, engaging with customers is a critical step in rebuilding trust and re-establishing a phoenix brand. The company needs to acknowledge the problem, listen to customers, take action, engage with customers, and rebuild trust. By following these steps, a company can successfully rise from the ashes and regain market trust.

Story Example:
Let's say that a real estate company called ABC Real Estate has been criticized for providing poor customer service and not following through on promises to its clients. This has resulted in a loss of market trust and a decrease in sales.

To rebuild trust with its customers, ABC Real Estate takes the following steps:

Acknowledge the problem: The company publicly acknowledges the issue and issues a public apology to its customers. It also sends out an email to its customer database, acknowledging the problem and expressing its commitment to improving its customer service.

Listen to customers: ABC Real Estate sends out a survey to its customers, asking for feedback on its customer service and ways it can improve. It also sets up a dedicated customer support hotline, where customers can call and speak with a representative directly.

Take action: Based on the feedback received from customers, ABC Real Estate takes action to address the issues. It hires additional customer service representatives to reduce wait times and improves its communication with customers. The company also implements a new policy where it follows up with clients after the sale is complete to ensure their satisfaction.

Engage with customers: ABC Real Estate starts a social media campaign to engage with its customers. It encourages customers to share their positive experiences with the company using the hashtag #ABCrealestate. The company also sends out regular email newsletters to its customer database, sharing news and updates about the company and its services.

Rebuild trust: ABC Real Estate offers incentives to encourage customers to give it another chance. It offers a discount on its services for customers who had previously experienced issues with the company. It also starts a referral program where customers can earn rewards for referring new business to the company.

Over time, ABC Real Estate's efforts to engage with customers and rebuild trust pay off. Customers start to see improvements in the company's customer service, and the company's reputation improves. Sales start to increase, and the company is able to regain its position as a trusted real estate brand in the market.

9

Chapter 9: Leveraging Influencers and Advocates

Chapter 9 of a marketing strategy book titled "Leveraging Influencers and Advocates" is focused on how companies can use influential individuals and brand advocates to improve their reputation and regain the trust of their target audience after facing criticism or negative publicity.

When a brand experiences a crisis that damages their reputation, it can be challenging to recover. The key to success is to take swift and decisive action to address the issue and communicate openly and honestly with your customers.

In the case of a phoenix brand, which has risen from the ashes after a crisis, it is essential to understand the power of influencers and advocates in rebuilding trust and credibility. These individuals have a loyal following and a significant impact on the purchasing decisions of their followers.

Here are some strategies that a phoenix brand can use to leverage influencers and advocates:

Identify and engage with relevant influencers: Influencers can help to amplify your message and rebuild your brand's reputation. Identify influencers who have a strong following in your industry or niche and reach out to them to establish a relationship. Offer them a product or service in exchange for their endorsement, or collaborate with them on a sponsored post or campaign.

Encourage user-generated content: Encourage your customers to share their positive experiences with your brand on social media and other platforms. You can incentivize this by offering a discount or free product to those who share their experiences. This will not only help to rebuild trust but also create a sense of community around your brand.

Leverage brand advocates: Brand advocates are customers who are extremely satisfied with your product or service and are willing to recommend it to others. Identify these advocates and reach out to them to see if they would be willing to share their experiences publicly. You can also offer incentives such as exclusive access to new products or services or early access to sales and promotions.

Be transparent and honest: It is crucial to be transparent and honest about the crisis that caused the damage to your brand's reputation. Acknowledge the issue and communicate what you are doing to address it. This will help to build trust and credibility with your target audience.

In summary, leveraging influencers and advocates is a powerful strategy for a phoenix brand to rebuild trust and credibility. By identifying relevant influencers, encouraging user-generated content, leveraging brand advocates, and being transparent and honest, a phoenix brand can recover from a crisis and regain the trust of their target audience.

Story Example:
Once upon a time, there was a young couple, Sarah and Alex, who had been dreaming of buying their first home. They had been searching for months, but everything they saw was either too small, too old, or too far from their workplaces.

One day, they stumbled upon a real estate brand,

let's call it "Dream Homes," that promised to help them find their dream home, no matter how elusive it seemed. The couple was hesitant at first, as they had been disappointed by other real estate agencies before. However, they decided to give it a try and contacted Dream Homes.

To their surprise, the Dream Homes team was incredibly responsive and genuinely interested in helping them find the perfect home. They asked them about their lifestyle, preferences, and budget, and used that information to suggest several properties that matched their criteria.

Sarah and Alex were thrilled to find that one of the properties was a perfect match for their needs. It was a charming, newly renovated house with a large backyard and a cozy fireplace, just a short drive away from their workplaces.

The Dream Homes team helped them navigate the complex home buying process, providing them with valuable insights and advice every step of the way. They also recommended a trusted mortgage broker who helped them secure a favorable loan, saving them thousands of dollars in interest rates.

Thanks to Dream Homes' dedication and expertise,

Sarah and Alex were able to buy their dream home stress-free, without any last-minute surprises or hidden fees. They moved in and started to make the house their own, feeling grateful for the support and guidance they received from Dream Homes.

From that day on, Sarah and Alex became loyal customers and advocates of Dream Homes. They recommended the real estate brand to all their friends and family, praising their professionalism, transparency, and genuine care for their clients. And, as they continued to enjoy their beautiful home, they knew they had made the right choice by trusting Dream Homes.

10

Chapter 10: Creating a Positive Customer Experience

Creating a positive customer experience is crucial for any brand, especially for a brand that has faced criticism and lost market trust. In this scenario, it is essential to take immediate steps to rebuild the brand's image and regain the trust of customers.

One approach to this is to use the concept of a phoenix brand, which means creating a new brand image from the ashes of the old one. This approach involves a complete overhaul of the brand's image, which includes everything from its messaging, marketing, and customer service to its product offerings and company culture.

Here are some steps that a brand can take to create a positive customer experience and regain

trust after facing criticism:

Acknowledge the problem: The first step in rebuilding a brand's image is to acknowledge the problem openly and honestly. The brand should take responsibility for its actions and show genuine concern for its customers' grievances.

Take action: The brand should take immediate action to address the issues that led to the criticism. This may involve revising the company's policies, improving its products or services, or implementing new customer service initiatives.

Communicate with customers: The brand should be transparent and communicative with its customers. This can be done through social media, email campaigns, or other forms of direct communication. By keeping customers informed, the brand can show that it is taking steps to address their concerns and regain their trust.

Offer incentives: To win back customers, the brand may need to offer incentives such as discounts, free trials, or other perks. This can show customers that the brand is committed to making things right and that it values their loyalty.

Focus on the customer experience: Finally, the brand should prioritize the customer experience in everything it does. This means creating a customer-centric culture that puts the needs and wants of customers first. By doing so, the brand can build long-lasting relationships with its customers and create a positive reputation in the market.

In conclusion, rebuilding a brand's image and regaining customer trust after facing criticism is a challenging task. However, by following the steps outlined above and using the phoenix brand approach, a brand can create a positive customer experience and rebuild its reputation. The key is to take immediate action, communicate with customers, and prioritize their needs and wants in everything the brand does.

Story Example:
Once a popular real estate brand, Cityscape Properties, was facing a significant setback. Several of their clients had complained about their lack of transparency in property deals and the poor quality of their customer service. The brand had lost market trust, and it was essential to take immediate action to rebuild its image.

To do this, Cityscape Properties decided to take a bold step and rebrand itself as New Horizons

Realty. They launched a new marketing campaign, emphasizing their commitment to transparency, honesty, and quality customer service. They also revamped their company culture, focusing on putting customers' needs and wants first.

New Horizons Realty began communicating regularly with their clients, sending out newsletters and emails to keep them informed about the latest developments in the real estate market. They also offered incentives such as discounts and free consultation services to win back customers.

To further enhance the customer experience, New Horizons Realty improved their website's user interface, making it easier for clients to find and browse properties. They also introduced a new mobile app, allowing clients to schedule property visits and track the progress of their deals.

All of these efforts paid off, and New Horizons Realty began to regain its market trust. Clients who had previously complained about the brand's lack of transparency and customer service began to notice significant improvements. They appreciated the brand's commitment to creating a positive customer experience and felt that their needs and wants were being prioritized.

In conclusion, New Horizons Realty's phoenix brand approach was a success, and it enabled them to create a positive customer experience and regain their market trust. By acknowledging their shortcomings, taking immediate action, and prioritizing their customers' needs and wants, they were able to rebuild their image and create a more positive reputation in the real estate market.

11

Chapter 11: Focusing on Product Quality

Chapter 11: Focusing on Product Quality for a real estate Phoenix brand which has lost its market share and is getting bad comments or reviews for its brand is an important chapter for the company to understand how they can regain their market share by improving the quality of their products.

The real estate industry is highly competitive, and companies need to ensure that they are delivering high-quality products to remain competitive. However, in some cases, companies may lose their market share and receive bad reviews due to a lack of focus on product quality. This can have a significant impact on the company's bottom line, as customers may choose to do business with competitors who are offering higher quality products.

In this chapter, the real estate Phoenix brand needs to understand the importance of focusing on product quality and how to go about improving it. The first step is to conduct a thorough analysis of their current products and identify areas where improvements can be made. This could include everything from the quality of construction materials to the design of the properties.

Once the company has identified areas for improvement, they need to take action to address these issues. This may involve investing in better materials or hiring a team of experts to improve the design of their properties. It is also important for the company to communicate these improvements to their customers, as this will help to rebuild trust and improve their reputation.

Another key aspect of focusing on product quality is ensuring that the company is delivering what their customers want. This means taking the time to listen to customer feedback and incorporating this feedback into future product designs. By doing this, the company can ensure that their products are meeting the needs of their target audience and are more likely to be well-received.

Finally, the company needs to ensure that they have a

strong quality control process in place to ensure that their products meet the highest standards. This may involve conducting regular inspections of their properties or hiring a third-party inspection company to review their work. By doing this, the company can ensure that their products are of the highest quality and are less likely to receive negative reviews.

In conclusion, Chapter 11: Focusing on Product Quality for a real estate Phoenix brand which has lost its market share and is getting bad comments or reviews for its brand is an important chapter for the company to understand how to regain its market share by improving the quality of its products. By taking a proactive approach to improving product quality, the company can rebuild trust with its customers, improve its reputation, and remain competitive in the highly competitive real estate industry.

Story Example:
There was a real estate brand based in Miami, Florida, that had been in business for over two decades. The brand had a strong reputation for delivering high-quality properties, and its customer base was primarily made up of high-net-worth individuals looking for luxurious homes and condos.

However, in recent years, the company had started to lose its market share. Customers were complaining about the quality of their properties, and the brand was receiving bad reviews online. The company's leadership team knew that they needed to take action to address these issues, or risk losing even more customers to their competitors.

The first step was to conduct a thorough analysis of their current properties and identify areas where improvements could be made. They found that many of their properties were outdated and not meeting the needs of their target audience. The company's leadership team realized that they needed to invest in better materials and work with architects and designers to improve the design of their properties.

Once they identified areas for improvement, the company took action to address these issues. They started investing in high-quality materials and working with renowned architects and designers to create new properties that met the needs of their target audience. The company also started to listen to customer feedback more closely and incorporate this feedback into their future designs.

The company then launched a marketing campaign to communicate these improvements to their customers. They shared before-and-after photos of their properties on social media and their website, highlighting the improvements they had made. They also started to offer new and improved amenities to their residents, such as rooftop pools, fitness centers, and private cinemas.

Over time, the company's reputation started to improve, and they started to receive positive reviews from their customers. Their market share began to increase, and they started to win back customers that they had lost to their competitors.

In conclusion, this real estate brand faced similar challenges as the Phoenix brand in Chapter 11. By focusing on product quality and investing in better materials and designs, the company was able to overcome these challenges and regain its market share. The key takeaway is that product quality is crucial for any business, and investing in it can pay off in the long run by improving customer satisfaction and driving business growth.

12

Chapter 12: Monitoring Online Reputation

Online reputation management is essential for businesses in today's digital age, and real estate brands are no exception. A brand's online reputation can significantly impact its success and its ability to attract new customers. In this chapter, we will discuss the importance of monitoring online reputation for a real estate Phoenix brand that has lost its market share and is receiving negative comments and reviews.

The first step in monitoring online reputation is to conduct a thorough analysis of the brand's current online presence. This includes identifying all the online platforms where the brand is mentioned, such as social media platforms, review sites, and business directories. Once these platforms are identified, the brand can begin to monitor the comments, reviews, and mentions that it receives on each platform.

In the case of the real estate Phoenix brand that has lost its market share and is receiving negative comments and reviews, it is essential to address these issues head-on. This means responding to negative comments and reviews in a timely and professional manner. By doing so, the brand can demonstrate its commitment to customer service and its willingness to address customer concerns.

When responding to negative comments and reviews, it is important to remain calm and professional. The brand should acknowledge the customer's concerns, apologize for any negative experiences, and offer a solution or a way to make things right. It is also essential to avoid getting into arguments or becoming defensive, as this can make the situation worse.

In addition to responding to negative comments and reviews, the real estate Phoenix brand can also take proactive measures to improve its online reputation. This includes regularly monitoring its online presence, creating high-quality content that showcases its expertise and commitment to customer service, and engaging with customers on social media platforms.

Another important aspect of monitoring online reputation is tracking the success of these efforts. This means tracking the brand's online engagement, monitoring changes in its online reputation, and regularly analyzing the effectiveness of its online reputation management strategies. By doing so, the brand can continually refine its approach and ensure that it is effectively managing its online reputation.

In conclusion, monitoring online reputation is crucial for real estate brands, particularly those that have lost market share and are receiving negative comments and reviews. By taking a proactive approach to online reputation management, responding to negative comments and reviews, and tracking the success of these efforts, the real estate Phoenix brand can improve its online reputation and rebuild its market share.

Story Example:
Once upon a time, there was a real estate brand in Phoenix that had been a market leader for years. However, with the rise of new competitors and changing consumer preferences, the brand started to lose its market share.

As the brand's market share dwindled, its online reputation also started to suffer. Customers were leaving negative comments and reviews on various online platforms, such as social media and review sites. The brand's reputation was taking a hit, and it seemed like it was losing its credibility in the eyes of potential customers.

The brand's management realized the importance of monitoring online reputation and decided to take proactive measures to address the issue. They hired a team of online reputation management experts to help them monitor the brand's online presence and respond to negative comments and reviews in a timely and professional manner.

The online reputation management team started by conducting a thorough analysis of the brand's current online presence. They identified all the platforms where the brand was mentioned and monitored the comments, reviews, and mentions that it received on each platform.

The team also started responding to negative comments and reviews in a timely and professional manner. They acknowledged the customer's concerns, apologized for any negative experiences, and offered solutions or ways to make things right. By doing so, they were able to demonstrate the brand's commitment to customer service and willingness to address customer concerns.

In addition to responding to negative comments and reviews, the team also took proactive measures to improve the brand's online reputation. They created high-quality content that showcased the brand's expertise and commitment to customer service. They also engaged with customers on social media platforms, responding to comments and addressing any concerns that customers may have had.

Over time, the brand's online reputation started to improve. The negative comments and reviews were becoming fewer and farther between, and the brand's credibility was being restored in the eyes of potential customers. The brand's market share also started to recover, and it was regaining its position as a market leader in the real estate industry in Phoenix.

In conclusion, the story of this real estate brand in Phoenix highlights the importance of monitoring online reputation for businesses in today's digital age. By taking a proactive approach to online reputation management and responding to negative comments and reviews in a timely and professional manner, businesses can improve their online reputation and rebuild their market share.

13

Chapter 13: Leveraging PR Strategies

In the competitive world of real estate, a brand's reputation is everything. A negative reputation can quickly diminish a brand's market share and ultimately lead to its downfall. If a real estate Phoenix brand has lost its market share and is receiving bad comments or reviews, it's important to take immediate action to address the situation. One effective approach to restoring a brand's reputation is to leverage public relations (PR) strategies.

Here are some steps that a real estate Phoenix brand can take to leverage PR strategies:

Conduct a Reputation Audit: The first step in any PR campaign is to conduct a thorough reputation audit. This includes reviewing online reviews, social media comments, and other feedback to understand what's causing the negative sentiment. It's important to understand the root cause of the negative reputation before developing a PR strategy.

Define the Target Audience: The next step is to define the target audience for the PR campaign. Is the negative reputation affecting potential home buyers, current residents, or other stakeholders? The messaging and tactics used in the PR campaign will depend on the target audience.

Develop a Messaging Strategy: Once the target audience is defined, the next step is to develop a messaging strategy. The messaging should be clear, concise, and focused on addressing the concerns raised in the reputation audit. It should also be consistent across all communication channels, including social media, press releases, and other PR materials.

Engage with Influencers: One effective PR strategy is to engage with influencers in the real estate industry. This includes bloggers, journalists, and social media influencers who have a large following in the target audience. By building relationships with these influencers, the real estate Phoenix brand can leverage their reach to amplify positive messaging and counter negative sentiment.

Host Events and Webinars: Hosting events and webinars is another effective PR strategy. These events can be used to showcase the brand's expertise and educate the target audience on relevant topics. By providing value to the target audience, the brand can build trust and credibility.

Monitor and Measure Results: Finally, it's important to monitor and measure the results of the PR campaign. This includes tracking online sentiment, engagement metrics, and other key performance indicators. By monitoring the results, the real estate Phoenix brand can adjust its strategy as needed to ensure the campaign is having the desired impact.

In conclusion, if a real estate Phoenix brand has lost its market share and is receiving negative comments or reviews, leveraging PR strategies can be an effective way to restore its reputation. By conducting a reputation audit, defining the target audience, developing a messaging strategy, engaging with influencers, hosting events and webinars, and monitoring results, the brand can turn its negative reputation around and regain market share.

Story Example:

Once upon a time, there was a real estate brand in Phoenix that had been in business for over two decades. The company had built a reputation for quality homes and excellent customer service, and it had a loyal following of homeowners who loved their homes.

However, as time went on, the real estate market in Phoenix became more competitive. Newer, flashier brands started to emerge, and the older real estate brand began to lose its market share.

At first, the company's management didn't pay much attention to this. They assumed that their reputation would carry them through, and that customers would continue to choose their homes over those of their competitors.

However, things started to change when negative comments and reviews started to appear online. Customers complained about slow response times, poor customer service, and quality issues with their homes. These negative reviews were starting to impact the company's reputation and its bottom line.

Realizing that they needed to take action, the company's management team decided to leverage PR strategies to turn things around. They started by conducting a reputation audit, reviewing all of the negative comments and reviews that had been posted online.

From there, they defined their target audience - potential home buyers who were considering purchasing a home in Phoenix. They developed a messaging strategy that focused on their commitment to quality and customer service, and they started engaging with influencers in the real estate industry to amplify their message.

To build trust and credibility with potential home buyers, the real estate brand also started hosting events and webinars. These events were designed to educate potential buyers on the home buying process and to showcase the company's expertise.

Finally, the real estate brand monitored and measured the results of their PR campaign. They tracked online sentiment, engagement metrics, and other key performance indicators to ensure that their messaging was having the desired impact.

Over time, the real estate brand's reputation began to improve. Positive reviews started to appear online, and customers began to give the company another chance. By leveraging PR strategies, the real estate brand was able to turn its negative reputation around and regain its market share in Phoenix.

14

Chapter 14: Harnessing the Power of SEO

Chapter 14 of the book "Harnessing the Power of SEO" is all about how a real estate brand in Phoenix can use search engine optimization (SEO) to regain its market share and improve its brand reputation. If a real estate brand has been getting bad comments or reviews and losing market share, SEO can help them recover by improving their online presence and visibility.

Here are some key steps that a real estate brand can take to harness the power of SEO and improve its online reputation:

Conduct a thorough audit of the website: The first step is to analyze the website and identify any issues that may be hindering its SEO performance. This could include slow loading times, broken links, duplicate content, or poor user experience. Once these issues are identified, they should be addressed as soon as possible.

Optimize website content: After identifying the issues, the website content should be optimized to make it more search engine friendly. This could include adding relevant keywords, creating engaging content, using header tags, and including internal and external links.

Claim and optimize local listings: Local listings such as Google My Business, Yelp, and Zillow can be incredibly valuable for real estate brands, especially those targeting a specific geographic area like Phoenix. By claiming and optimizing these listings, brands can improve their visibility in local search results and attract more potential customers.

Build quality backlinks: Backlinks are an important factor in SEO and can help improve a brand's reputation and authority. Building high-quality backlinks from reputable websites can help boost the website's ranking in search results and improve its overall visibility.

Monitor and respond to online reviews: Online

reviews can have a significant impact on a real estate brand's reputation, so it's important to monitor them regularly and respond to any negative feedback. Responding to negative feedback in a professional and constructive manner can help improve the brand's image and demonstrate a commitment to customer service.

By following these steps, a real estate brand in Phoenix can harness the power of SEO to improve its online reputation and regain its market share. It's important to remember that SEO is a long-term strategy and requires ongoing effort and investment to see results. However, with patience and persistence, it's possible to achieve significant improvements in search engine rankings and overall brand visibility.

Story Example:
Let's say there's a real estate brand called "Phoenix Homes" that has been in business for many years, but in recent times, they have been struggling to maintain their market share due to the rise of new competitors and negative online reviews from dissatisfied customers.

The management team of Phoenix Homes decides to invest in SEO to improve their online presence and reputation. They hire a digital marketing agency to help them with the process. The agency begins by conducting a thorough audit of the Phoenix Homes website and identifies several issues that are hindering its SEO performance.

The agency works with the Phoenix Homes team to optimize the website content by adding relevant keywords, creating engaging content, and improving the user experience. They also claim and optimize local listings on Google My Business, Yelp, and Zillow to improve their visibility in local search results.

To improve the brand's authority and reputation, the agency works on building high-quality backlinks from reputable websites in the real estate industry. They also monitor and respond to online reviews regularly, demonstrating a commitment to customer service and improving the brand's image.

Over time, Phoenix Homes' search engine rankings improve significantly, and they begin to attract more traffic to their website. Customers start to notice the improvements in the brand's online reputation and begin to leave more positive reviews.

As a result of the investment in SEO, Phoenix Homes

begins to regain its market share in the real estate industry in Phoenix. The brand's online presence improves, and they are now recognized as a reliable and trustworthy real estate company by potential customers.

In conclusion, the story of Phoenix Homes illustrates how investing in SEO can help a struggling real estate brand regain its market share and improve its online reputation. By following the steps outlined in Chapter 14 of "Harnessing the Power of SEO," real estate brands can overcome negative online reviews and improve their visibility in search engine results.

15

Chapter 15: Developing a Crisis Management Plan

Crisis management is a crucial aspect of any business, especially when it comes to real estate brands that have lost their market share and are facing negative reviews or comments from customers. In this scenario, it becomes essential to develop a crisis management plan that can help the brand to recover from the situation and regain its lost reputation.

Chapter 15 of a crisis management plan for a real estate brand that has lost its market share and is facing negative reviews and comments should cover the following aspects:

Assess the situation: The first step in developing a crisis management plan is to assess the situation. This involves identifying the cause of the crisis, the extent of the damage, and the potential impact it may have on the brand's reputation. It is important to gather as much information as possible to make informed decisions.

Identify key stakeholders: The next step is to identify key stakeholders who will be affected by the crisis. This includes customers, employees, shareholders, and the media. The crisis management plan should address the concerns of each stakeholder group and provide a clear communication strategy to keep them informed.

Develop a crisis communication strategy: Communication is key during a crisis. The crisis management plan should include a communication strategy that outlines how the brand will communicate with stakeholders, what messages will be conveyed, and who will be responsible for delivering those messages. It is important to be transparent and honest in communication to build trust and credibility.

Take action: The crisis management plan should include a plan of action to address the root cause of the crisis. This may involve implementing changes in the brand's operations, addressing customer concerns, and taking steps to improve the brand's reputation.

Monitor and evaluate: It is important to continuously monitor the situation and evaluate the effectiveness of the crisis management plan. This will help to identify any gaps or areas for improvement and make necessary adjustments to the plan.

Learn from the crisis: Once the crisis has been resolved, it is important to reflect on the lessons learned and incorporate them into the brand's operations. This will help to prevent similar crises from occurring in the future.

In conclusion, a crisis management plan is essential for real estate brands that have lost their market share and are facing negative comments or reviews. The plan should include a clear assessment of the situation, identification of key stakeholders, a communication strategy, a plan of action, monitoring and evaluation, and learning from the crisis. By following these steps, the brand can recover from the crisis and regain its reputation.

Story Example:
Imagine a real estate brand called "Dream Homes" that has been a well-established player in the market for years. However, due to the economic downturn and increased competition, the brand has been losing its market share. This has resulted in negative

comments and reviews from customers who are dissatisfied with the quality of services and the high prices of Dream Homes' properties.

The situation has worsened to a point where customers have started to boycott Dream Homes' properties, and the media has started to report negative stories about the brand. The management team at Dream Homes realizes that they need to take immediate action to save the brand's reputation and regain customers' trust.

They decide to develop a crisis management plan, and the first step is to assess the situation. They conduct a thorough analysis of the causes of the crisis and the extent of the damage to the brand's reputation. They identify the key stakeholders who are affected by the crisis, including customers, employees, shareholders, and the media.

Based on the assessment, the management team develops a crisis communication strategy. They decide to be transparent and honest in their communication, acknowledging the issues and concerns raised by customers and the media. They assign a team to handle communications with the stakeholders, ensuring that they keep everyone informed of the actions being taken to address the crisis.

The management team takes action to address the root cause of the crisis. They conduct a comprehensive review of their operations, identifying areas where they can improve the quality of their services and reduce the prices of their properties. They also implement changes in their marketing strategies, targeting new demographics and offering incentives to existing customers.

As part of their crisis management plan, the management team monitors and evaluates the situation closely, making adjustments as necessary to the plan. They track the progress of the brand's reputation, and the impact of the changes made in their operations and marketing strategies.

In the end, the crisis management plan proves to be successful. Customers start to notice the changes made by Dream Homes, and the brand's reputation starts to improve. The media reports on the positive steps taken by the brand, and customers begin to return to Dream Homes' properties. The management team learns from the crisis and incorporates the lessons into the brand's operations, ensuring that they are better prepared for any future challenges.

16

Chapter 16: Building a Positive Corporate Culture

For a real estate brand that has lost its market share and is receiving negative comments or reviews, it is crucial to focus on building a positive corporate culture. A positive corporate culture is a fundamental aspect of any successful business, and it can have a significant impact on a brand's reputation and customer perception.

Here are some steps that the real estate brand can take to build a positive corporate culture:

- **Define your brand values and mission**

The first step in building a positive corporate culture

is to define your brand values and mission. This will help establish the core principles that your brand stands for and the vision that you aim to achieve. Your brand values should be communicated to employees and stakeholders, and they should be reflected in every aspect of your business, including your marketing and branding efforts.

- **Hire the right people**

A positive corporate culture begins with the people you hire. When hiring new employees, look for candidates who share your brand values and are committed to delivering exceptional customer service. Additionally, it is essential to create a diverse and inclusive workplace where everyone feels valued and respected.

- **Invest in employee training and development**

Investing in employee training and development is essential to building a positive corporate culture. Provide your employees with the tools and resources they need to succeed, and encourage them to take ownership of their roles. Additionally, offer ongoing training opportunities to help them grow and develop professionally.

- **Encourage collaboration and communication**

Encouraging collaboration and communication is crucial to building a positive corporate culture. Encourage teamwork and open communication across all levels of your organization. This will help foster a sense of community and create a supportive work environment.

- **Recognize and reward success**

Recognizing and rewarding success is essential to building a positive corporate culture. Celebrate your employees' achievements and encourage them to take pride in their work. Additionally, offer incentives and rewards for outstanding performance to encourage continued excellence.

- **Embrace transparency and accountability**

Embracing transparency and accountability is crucial to building a positive corporate culture. Be open and honest with your employees and stakeholders, and hold yourself and your team accountable for your actions. This will help build trust and credibility and demonstrate your commitment to ethical business practices.

In conclusion, building a positive corporate culture is essential for any business, particularly a real estate brand that has lost its market share and is receiving negative comments or reviews. By defining your brand values and mission, hiring the right people, investing in employee training and development, encouraging collaboration and communication, recognizing and rewarding success, and embracing transparency and accountability, you can create a supportive work environment that fosters success and helps rebuild your brand's reputation.

Story Example:

Imagine a real estate brand called "Sunset Properties," which was once a market leader in the city but has lost its market share in recent years. Customers have been leaving negative comments and reviews online, complaining about poor customer service and outdated marketing strategies.

To address these issues, Sunset Properties decided to focus on building a positive corporate culture. The company's leadership team began by defining the brand's values and mission, which centered on providing exceptional customer service and delivering on the promise of finding customers their dream homes.

Next, Sunset Properties revamped its hiring process, seeking out candidates who shared the brand's values and were committed to delivering exceptional customer service. The company also invested heavily in employee training and development, providing ongoing opportunities for professional growth and development.

Sunset Properties also encouraged collaboration and communication among its employees, creating a supportive work environment that fostered teamwork and open communication. The company recognized and rewarded success, celebrating its employees' achievements and incentivizing outstanding performance.

Finally, Sunset Properties embraced transparency and accountability, being open and honest with its employees and stakeholders, and holding itself and its team accountable for its actions.

As a result of these efforts, Sunset Properties began to see a significant improvement in its customer satisfaction ratings and reviews. Customers praised the company's exceptional customer service, innovative marketing strategies, and commitment to ethical business practices.

Over time, Sunset Properties regained its market share, becoming a market leader once again. By focusing on building a positive corporate culture, the brand was able to rebuild its reputation and regain its customers' trust and loyalty.

Chapter 17: Encouraging Customer Loyalty

Chapter 17: Encouraging Customer Loyalty is an important topic for any business, including a real estate brand that has lost its market share and is receiving negative feedback from its customers. In this chapter, we will discuss some strategies that can be used to encourage customer loyalty and improve the reputation of a real estate brand.

Address the issues: The first step towards encouraging customer loyalty is to address the issues that have caused the negative feedback. Conduct surveys, listen to customer complaints, and take prompt action to resolve the problems. Addressing the issues shows that you value your customers and are committed to providing them with the best service.

Improve communication: Effective communication is key to building customer loyalty. Keep your customers informed about the progress of their transactions and provide regular updates. Respond promptly to emails, phone calls, and messages. Make sure that your customers are aware of any changes in the market, property prices, or other relevant information.

Offer incentives: Offering incentives is a great way to encourage customer loyalty. You can offer discounts, special promotions, or loyalty programs to reward your customers for their business. This not only helps to retain existing customers but also attracts new ones.

Provide excellent customer service: Providing excellent customer service is essential for any business. Make sure that your staff is well-trained and equipped to handle customer queries and complaints. Responding to complaints in a timely and professional manner can turn a dissatisfied customer into a loyal one.

Build relationships: Building relationships with your customers is another effective way to encourage loyalty. Get to know your customers personally and take an interest in their lives. This can be done by sending personalized messages, wishing them on special occasions, or inviting them to events. Building relationships helps to create a sense of trust and loyalty.

Maintain a positive online presence: In today's digital age, it's essential to maintain a positive online presence. Respond to negative reviews and comments promptly and professionally. Highlight positive feedback and reviews on your website and social media platforms. This shows that you care about your customers and are committed to providing them with the best service.

In conclusion, encouraging customer loyalty is essential for any business, especially a real estate brand that has lost its market share and is receiving negative feedback. By addressing issues, improving communication, offering incentives, providing excellent customer service, building relationships, and maintaining a positive online presence, you can encourage customer loyalty and improve your brand's reputation.

Story Example:
Imagine a real estate brand called "Dream Homes" that has been in the industry for over a decade. They were once one of the leading real estate agencies in

the region, with a reputation for providing excellent customer service and delivering on promises. However, in recent years, their market share has been dwindling, and they have been receiving negative reviews and comments from dissatisfied customers.

One of their clients, Emily, had a terrible experience with Dream Homes. She had been searching for her dream home for months and finally found a property listed by Dream Homes. She called their agent, who promised to take her on a tour of the property. However, when Emily arrived at the property, the agent was nowhere to be found. She tried calling him several times, but he did not answer. Frustrated and disappointed, Emily left the property and decided not to deal with Dream Homes again.

To address their declining market share and negative feedback, Dream Homes took the following steps:

Addressed the issue: The management team of Dream Homes conducted a review of their customer service practices and identified areas that needed improvement. They then made the necessary changes to ensure that their agents were more responsive to their clients' needs.

Improved communication: Dream Homes implemented a system of regular updates for their clients, keeping them informed about the status of their property search, the progress of their transactions, and any relevant market information.

Offered incentives: Dream Homes introduced a loyalty program that offered discounts on future transactions and referral bonuses to existing customers.

Provided excellent customer service: Dream Homes invested in training their staff and equipped them with the necessary skills and tools to handle customer queries and complaints promptly and professionally.

Built relationships: Dream Homes introduced personalized messaging for their customers, wishing them on special occasions, and inviting them to events. They made an effort to get to know their clients personally and showed a genuine interest in their lives.

Maintained a positive online presence: Dream Homes responded to negative reviews and comments promptly and professionally, highlighting positive feedback and reviews on their website and social media platforms.

These measures proved successful in improving Dream Homes' reputation and customer loyalty. Emily, the dissatisfied customer, gave Dream Homes another chance, and this time, they delivered on their promises. Emily found her dream home with the help of a responsive and attentive Dream Homes agent. She was impressed with their efforts to address their customer service issues and even recommended Dream Homes to her friends and family.

In conclusion, Dream Homes' success story shows that by addressing issues, improving communication, offering incentives, providing excellent customer service, building relationships, and maintaining a positive online presence, a real estate brand can improve its reputation and encourage customer loyalty.

18

Chapter 18: Capitalizing on Social Media Opportunities

Chapter 18: Capitalizing on Social Media Opportunities for a real estate brand which has lost its market share and is being getting bad comments or reviews for its brand

Social media has become a crucial part of marketing strategies for businesses of all sizes, including real estate companies. In today's digital age, where people spend more time online, social media platforms provide an excellent opportunity for brands to connect with their audience and promote their products and services. However, if a real estate brand has lost its market share and is getting bad comments or reviews, it is essential to capitalize on social media opportunities to improve its brand image.

Here are some strategies that a real estate brand can implement to capitalize on social media opportunities:

Listen to Customer Feedback: The first step in improving a brand's image on social media is to listen to what customers are saying about the brand. Real estate companies can use social media monitoring tools to track their brand mentions and customer feedback. Once they have identified the areas where they are lacking, they can work on improving those aspects of their business.

Respond to Negative Comments: It is important for real estate companies to respond to negative comments on social media promptly. They should acknowledge the customer's concerns and offer a solution or an explanation for the issue. This shows that the company values its customers and is willing to address their concerns.

Engage with Customers: Social media provides an excellent opportunity for real estate companies to engage with their customers. They can use social media to answer customer queries, provide helpful tips and advice, and share valuable content related to the real estate industry. This helps to build trust and establish the company as an authority in the industry.

Create Valuable Content: Real estate companies can use social media to create valuable content that educates and informs their audience. They can share blog posts, videos, and infographics that provide useful information related to the real estate industry. This helps to establish the company as an expert in the industry and improves their brand image.

Run Social Media Campaigns: Real estate companies can run social media campaigns to promote their brand and increase their reach. They can use paid social media advertising to target their audience and promote their products and services. They can also run contests and giveaways to engage with their audience and generate buzz around their brand.

In conclusion, social media provides an excellent opportunity for real estate brands to improve their brand image and regain their market share. By listening to customer feedback, responding to negative comments, engaging with customers, creating valuable content, and running social media campaigns, real estate companies can capitalize on social media opportunities and establish themselves as a reputable brand in the industry.

Story Example:

Once a real estate brand named XYZ was a well-known and trusted name in the industry. However, over time, they began to lose their market share due to the rise of new competitors and negative feedback from some customers.

To turn things around, the company decided to capitalize on social media opportunities. They started by listening to customer feedback and identifying the areas where they were lacking. They realized that their customer service needed improvement, and they had to work on being more responsive to customer queries and concerns.

To address this, they created a social media team dedicated to monitoring their brand mentions and responding to customer feedback promptly. They also started engaging with customers on social media by providing helpful tips and advice related to the real estate industry.

In addition, they created valuable content such as blog posts, videos, and infographics to educate and inform their audience about the real estate market. They also ran social media campaigns to promote their brand and increase their reach.

One of their successful campaigns was a social media contest where participants had to share a photo of their dream home and tag XYZ's social media page. The winner would receive a free consultation with XYZ's real estate experts. The campaign generated a lot of buzz around the brand, and the company saw an increase in website traffic and sales leads.

Through these efforts, XYZ was able to improve their brand image on social media, establish themselves as an authority in the real estate industry, and regain their market share. They became known as a customer-centric and innovative brand, and their efforts were appreciated by their customers, resulting in positive feedback and reviews.

19

Chapter 19: Collaborating with Industry Leaders

Chapter 19: Collaborating with Industry Leaders for a real estate brand which has lost its market share and is getting bad comments or reviews for its brand

Introduction:

The real estate industry is highly competitive, and it can be challenging for brands to maintain their market share. A real estate brand that has lost its market share and is receiving negative comments or reviews can benefit greatly from collaborating with industry leaders. In this chapter, we will explore how collaborating with industry leaders can help such a brand improve its market share and reputation.

Step 1: Identifying the right industry leaders

The first step in collaborating with industry leaders is to identify the right people or organizations to collaborate with. The real estate brand should research the market and identify the individuals or organizations that are influential in the industry. These individuals or organizations could be real estate agents, property developers, or industry experts.

Step 2: Approaching the industry leaders

Once the real estate brand has identified the right industry leaders, the next step is to approach them. The brand should reach out to the industry leaders and explain their situation. They should highlight their brand's strengths and the areas in which they need improvement. The brand should also explain how they believe that collaborating with the industry leaders could benefit both parties.

Step 3: Working together

Once the real estate brand has successfully collaborated with the industry leaders, they should work together to develop strategies that will improve the brand's market share and reputation. The industry leaders could offer valuable insights into the market and provide the real estate brand with strategies that have worked for them in the past. The real estate brand should be open to new ideas and willing to make changes to improve their business.

Step 4: Monitoring the results

After implementing the strategies developed in collaboration with the industry leaders, the real estate brand should monitor the results. They should measure the success of the strategies and make adjustments as necessary. They should also continue to collaborate with industry leaders to develop new strategies and stay up-to-date with the latest trends in the market.

Conclusion:

Collaborating with industry leaders is an effective way for a real estate brand that has lost its market share and reputation to improve its standing in the market. By identifying the right industry leaders, approaching them, and working together to develop strategies, the real estate brand can improve its market share and reputation. It is essential to monitor the results and continue to collaborate with industry leaders to stay ahead of the competition.

Story Example:
Case Study: Rebuilding a Real Estate Brand with Industry Leaders

ABC Realty was a real estate brand that had been a dominant player in the market for decades. However, in recent years, the brand had lost its market share and reputation due to negative comments and reviews from customers. The company was struggling to attract new clients, and their existing clients were leaving for their competitors.

To rebuild the brand, ABC Realty decided to collaborate with industry leaders. The company identified several successful real estate agents and property developers in the area and approached them with a proposal to work together to improve the brand's market share and reputation.

The industry leaders were initially hesitant to work with ABC Realty, given its poor reputation in the market. However, ABC Realty explained its strengths and the areas in which it needed improvement. The company also highlighted the benefits of collaborating with the industry leaders, including the potential for increased business for all parties.

After initial discussions, the industry leaders agreed to collaborate with ABC Realty. Together, they developed a strategy that focused on improving the customer experience, providing value-added services, and leveraging technology to make the buying and selling process more efficient.

The industry leaders provided ABC Realty with valuable insights into the market and offered strategies that had worked for them in the past. They also shared their best practices and helped ABC Realty implement them in their business.

ABC Realty launched the new strategy with a fresh brand identity and a new marketing campaign. The company focused on providing excellent customer service and value-added services, such as free home staging and property inspections. The company also leveraged technology to make the buying and selling process more efficient, such as offering virtual home tours and digital property listings.

The results were remarkable. ABC Realty's market share increased by 25%, and the company's reputation improved significantly. The brand's net promoter score, which measures customer satisfaction and loyalty, also increased by 15 points.

Conclusion:

By collaborating with industry leaders, ABC Realty was able to rebuild its brand and improve its standing in the market. The company's willingness to be open to new ideas and make changes to its business model was crucial to its success. The collaboration with industry leaders provided the company with valuable insights into the market and offered strategies that had worked for other successful businesses. As a result, ABC Realty was able to differentiate itself from its competitors, attract new clients, and retain its existing ones.

20

Steps with timelines of how this can be done with estimated time which it will take!

When a real estate brand loses its market share and begins receiving negative comments or reviews, it is essential to take swift action to address the situation. Here are some steps that the brand can take, along with estimated timelines for each step:

Conduct a thorough review: The first step is to conduct a thorough review of the situation. This includes analyzing the market, understanding the reasons behind the brand's decline, and assessing the quality of the brand's products and services. This process can take anywhere from two weeks to a month, depending on the complexity of the situation.

Develop a strategy: Based on the findings of the review, the brand should develop a strategy to address the issues. This strategy should include steps to improve the quality of the brand's products and services, enhance its marketing and branding efforts, and address any customer complaints or negative reviews. This process can take two to three weeks.

Implement the strategy: Once the strategy has been developed, it's time to start implementing it. This can involve making changes to the brand's products and services, launching new marketing and advertising campaigns, and addressing any customer complaints or negative reviews. Depending on the scope of the strategy, this process can take anywhere from three months to a year.

Monitor progress: As the brand implements its strategy, it's important to monitor progress and track the results. This allows the brand to identify areas that are working well and areas that need further improvement. This process should be ongoing and can take anywhere from six months to two years.

Evaluate and adjust: After a period of time, the brand should evaluate the effectiveness of its strategy and make any necessary adjustments. This could involve refining marketing and branding efforts, improving customer service, or making further changes to the brand's products and services. This process can take several months.

Overall, it's important to remember that rebuilding a brand's market share and reputation takes time and effort. The estimated timelines provided above are only rough estimates, and the actual time required will depend on a variety of factors, including the severity of the brand's decline and the effectiveness of its strategy.

Story Example:
Once a successful real estate brand, "Cityscape Properties" had been losing its market share and receiving negative reviews from customers for some time. Customers were complaining about the quality of the properties, slow customer service, and lack of transparency in the buying process. The brand had to take immediate action to address the situation.

To begin with, Cityscape Properties conducted a comprehensive review of its operations. They analyzed the market and found that their competitors were offering better quality properties at competitive prices. Additionally, they found that the company's internal processes were not streamlined, leading to delays and confusion among customers. The review process took two weeks.

Based on their findings, Cityscape Properties developed a strategy to address the issues. They implemented a new quality control process to ensure that all properties met their high standards. They also introduced a new customer service team to handle complaints and provide more transparent communication with customers. In addition, they started an aggressive marketing campaign to promote their brand and properties.

Cityscape Properties implemented this strategy over the course of a year. They introduced new properties that met their high-quality standards and provided more detailed information about the buying process. They also focused on improving their online presence, including their website and social media platforms, to reach a wider audience.

Over the course of the next six months, Cityscape Properties monitored the progress of their strategy. They analyzed their customer feedback and found that customers were much happier with their service and the quality of their properties. In response to customer feedback, they continued to refine their customer service processes to ensure that all customer concerns were addressed quickly and effectively.

After two years, Cityscape Properties evaluated the effectiveness of their strategy and found that their market share had increased, and customer satisfaction had improved significantly. They continued to adjust their approach, focusing on areas where they could improve further.

Overall, Cityscape Properties demonstrated that with a strategic approach and a commitment to improvement, a real estate brand can recover from a decline in market share and rebuild its reputation.